MY HAPPINESS COUNTS

For my big brother Kareem who encouraged me to write and taught me that my happiness counts - C.W.D.

Happiness is everywhere, just waiting to be found.

I take a moment to look within,
and to look around.

to match mine.

as I cross the finish line.

3 special words in my family are

4 seasons to enjoy.

Each offering something new.

I sow the seeds then watch them sprout.

I use each one to make beautiful art.

10 friends join me to count again.

Help me show them the way!

THE AUTHOR

Charm Joy Der, MA Ed, PHR, SHRM-CP is a wife, a mother to two energetic and inquisitive boys, an entrepreneur, and a diversity, equity, and inclusion community educator. With her debut book, My Happiness Counts, Charm aims to help young readers see that happiness is within their reach and that their happiness matters. She also wants to ensure that diversity and inclusion are integral parts of children's literature. Originally from Detroit, Michigan she now resides in Wisconsin with her family.

Lisa Wee is an award winning illustrator. Her quaint, quirky and vibrant illustrations depict diversity and inclusivity. She has worked with numerous self published authors and publishers on creating picture books with topics ranging from girls in STEM to self care for children.

THE ILLUSTRATOR

All rights reserved. No part of this book may be reproduced, stored in a retrieval system, or transmitted in any form by any means without prior written permission from the copyright owner.

Text Copyright © 2021 by Charm Der
Illustrations Copyright © 2021 by Lisa Wee
ISBN 979-8-7284-3305-7 (paperback)

Published by With Wonder Publishing, LLC
www.myhappinesscounts.com

www.ingramcontent.com/pod-product-compliance
Lightning Source LLC
Chambersburg PA
CBHW051829210526
45473CB00005B/1798